**ISES**

International
Solar Energy
Society

# Wind Energy Pocket Reference

**Editor of Series:**

**D. Yogi Goswami**

**Peter H. Jensen    Niels I. Meyer**

**Niels G. Mortensen    Flemming Oster**

T0321415

Routledge
Taylor & Francis Group

LONDON AND NEW YORK

## PREFACE

The ISES *Wind Energy Pocket Reference* contains useful information and data for wind energy developers, energy planning authorities, designers, technicians and those generally interested in wind power systems and their potential in energy supply systems.

The pocket book contains data, equations and figures for global distribution of wind energy, wind turbine technologies, turbine energy efficiency, systems considerations, and annual energy production under different local conditions, organizational schemes and economy of wind power.

The information is gathered from published literature supplemented by personal information of experts from energy agencies, utilities, universities and other energy research institutions, and from the wind energy industry and different types of wind power producers. We wish to thank all who have contributed information to this pocket book. We also would like to thank Ms. Barbara J. Graham, Assistant Editor, whose hard work and dedication was essential in compiling this volume.

**Authors:** Peter H. Jensen, Ph.D.*, Niels I. Meyer, Ph.D.**, Niels G. Mortensen, Ph.D.* and Flemming Øster, Ph.D.*

*Wind Energy Department, Risø National Laboratory for Sustainable Energy, Technical University of Denmark.

** Dept. of Civil Engineering, Technical University of Denmark

**Editor:**

> D. Yogi Goswami, Ph.D., P.E.
> Clean Energy Research Center
> College of Engineering
> University of South Florida
> Tampa, USA

---

*The Wind Energy Pocket Reference is the second in a series of pocket books to be published by ISES. The first in the series was the Solar Energy Pocket Reference (2005).*

*We hope this series will serve the practitioners of the various renewable energy fields.*

**Torben Esbensen**
**ISES President**
**(2006-2007)**

---

First published 2007 by Earthscan

Published 2019 by Routledge
2 Park Square, Milton Park, Abingdon, Oxon OX14 4RN
52 Vanderbilt Avenue, New York, NY 10017

*Routledge is an imprint of the Taylor & Francis Group, an informa business*

ISBN 978-1-84407-539-3 (pbk)

The series includes the following topics:

Solar Energy Reference Pocket Book
© 2005 ISES   ISBN 0-9771282-0-2

Wind Energy Reference Pocket Book
© 2007 ISES   ISBN 0-9771282-1-0

Bio Energy Reference Pocket Book
© 2009 ISES   ISBN 0-9771282-2-9

# TABLE OF CONTENTS

## I. INTRODUCTION

In 1900, wind turbines were typically used for grinding, water pumping and other rural mechanical energy needs. At that time a Danish inventor Poul la Cour at Askov Højskole (Folk High School) made experiments with Dutch wind mills converting them to fast running electricity production turbines. La Cour constructed wind tunnels for the purpose of developing aerodynamically shaped rotor blades.

The wind turbine history is summarized in Table I.1 emphasizing the Danish case as an example. Electrification of rural Danish areas was the motivation behind of the first commercial round of electricity producing turbines. During the Second World War wind energy flowered again followed by post war development of the medium sized Danish Gedser turbine. This turbine was the model for the modern development following the oil crises in the 1970s.

Many studies have estimated the total global wind energy resources. These studies have confirmed that the world's wind resources are extremely large and well distributed across almost any region and country. Lack of resource is therefore unlikely to become a limiting factor in the utilization of wind power for electricity generation. Table 1.2 gives estimates of the global wind production potential (Rasmussen and Øster, 1990; BTM Consult ApS, 2007).

Table I.1: Wind Turbine History Focusing on Danish Turbines and Electricity Production

| Year | Technology | Application | Comments | Regulation |
|---|---|---|---|---|
| 1235 – 1900 | Wind roses, Dutch windmills | Mechanical energy in agriculture | Grinding mills, water pumping etc. | |
| 1890 – 1910 | Horizontal axis turbines with 3-4 blades | Electricity production | Pioneered by Danish Poul la Cour at Askov Folk High School. Ref. I.1. | Yawing control |
| 1910 – 1950 | Horizontal axis, 3 blades. Turbine sizes of up to 5 MW | Electricity production in local power plants | Further developments in Denmark, UK, USA and USSR. Eventually out-competed by cheap oil | Spoiler rail for control |
| 1950 – 1967 | Horizontal axis, 3 blades, 200 kW stall controlled Danish Gedser Wind Turbine | Electricity production, utility demonstration project | The Gedser Mill (inventor Johs Juul) became the mother of modern Danish wind turbines | Stall control |
| 1974 | Start of modern phase of wind power pioneered by Danish manufacturers and developers | | Increasing turbine capacity from 22 kW in 1975 (inventor Chr. Riisager) Increasing sizes up to 6.5 MW in 2007 | Stall or pitch control |
| | | Grid-connected electricity production | | |
| 1980 | Pitch controlled turbines | | | |
| 1991 | First offshore wind farm in the Danish part of the Baltic Sea | | Eleven 450 kW turbines about 2 km from the shore | Active stall control |

Table I.2: Global potential for wind power production (Centre for Renewable Energy Technologies, 2000; European Wind Energy Association and Greenpeace, 2004; Grubb and Meyer, 1993).

| Reference | Region | Potential Production | Comments |
|---|---|---|---|
| ETSU, Harwell, UK | Europe | 330 TWh/year | Conservative estimate |
| Grubb and Meyer, in Renewable Energy, 1993 | World | 53,000 TWh/year | More than twice the projected electricity demand in 2020 |
| Wind Force 12, by EWEA and Greenpeace, February 2004 | World | 53,000 TWh/year | Surprisingly the same result as earlier by Grubb and Meyer |

# 1. GLOBAL DEVELOPMENT OF WIND ENERGY

## 1a. Global development of capacity (onshore and offshore)

### 1a1. Installed total wind turbine capacity globally

By the end of 2006 more than 74 GW of wind power was installed in the world, of which 48.6 GW or 65% was in Europe. Fig. 1.1 presents the development in global installed wind turbine capacity from 1990 to 2006 and a forecast of the expected future global capacity for the next decade (Danish Energy Agency, 1990). Global total installed capacity rose by an average of more than 25% per year since 2000; the capacity has been increased more than five times since then.

Fig. 1.2 shows the distribution by continents of wind turbine capacity world-wide. Fig. 1.3 (a and b) shows the distribution of installed wind turbines by year and continent.

### 1a2. Installed capacity offshore

It is expected that an increasing share of the new wind energy capacity in the future will be placed in offshore wind farms, where wind resources compared to conditions on land generally are plentiful and the wind climate less complicated favoring a longer turbine life time. Also any visual inconveniences imposed by large onshore wind turbines are eliminated or reduced. Details are given in Chapter 5.

*Fig. 1.1: Installed wind power in the world: Development and forecast. (Ref I.2)*

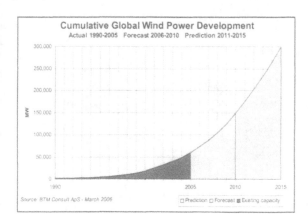

*Fig. 1.2: Distribution by continents of wind turbine capacity worldwide.*

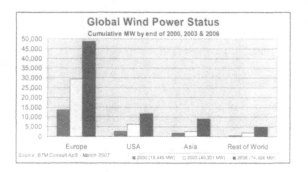

*Fig. 1.3: Distribution of installed wind turbines by year (a) and continent (b).*

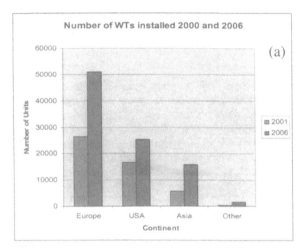

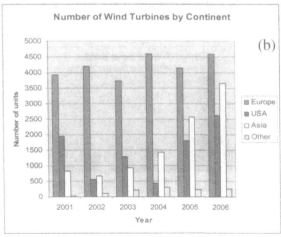

## 1b. Onshore and offshore markets for wind turbines

Within European countries, the amount of wind-generated electricity in 2006 measured as a fraction of the electricity demand varies from a barely detectable amount to about 17% in Denmark (BTM Consult ApS, 2007). Germany, Spain and Portugal are next on the scale with about 8%, 7%, and 6% of the demand produced by wind (Table 1.1).

Table 1.1: Largest 10 markets in 2006 and their development from 2003.

| Market | 2003, GW | 2004, GW | 2005, GW | 2006 GW | 2006, % |
|---|---|---|---|---|---|
| USA | 1.7 | 0.4 | 2.4 | 2.5 | 16.3 |
| Germany | 2.7 | 2.1 | 1.8 | 2.2 | 14.9 |
| India | 0.4 | 0.9 | 1.4 | 1.8 | 12.3 |
| Spain | 1.4 | 2.1 | 1.8 | 1.6 | 10.6 |
| PR China | 0.1 | 0.2 | 0.5 | 1.3 | 8.9 |
| France | 0.1 | 0.1 | 0.4 | 0.8 | 5.4 |
| Canada | - | 0.1 | 0.2 | 0.8 | 5.2 |
| UK | 0.2 | 0.3 | 0.5 | 0.6 | 4.2 |
| Portugal | 0.1 | 0.3 | 0.5 | 0.6 | 4.2 |
| Italy | 0.1 | 0.4 | 0.5 | 0.4 | 2.8 |
| Total | 6.8 | 6.7 | 9.9 | 12.7 | |
| % of word market | 81.7 | 82.5 | 85.9 | 84.7 | |

Source: World Market Update, BTM Consult ApS (March 2006).

## 2. WIND RESOURCES

### 2a. Wind climate and wind energy

The power density $E$ in the wind, which is equal to the kinetic energy flux per unit area perpendicular to the flow, is given by:

$$E = \frac{1}{2}\rho V^3$$

$\rho$ is the air density and $V$ the wind speed. When wind energy is converted to electricity by means of a wind turbine the power output is given by:

$$P = \frac{1}{2}\rho V^3 A C_p$$

where $P$ is power output (in Watts), $A$ the rotor area and $C_p$ the efficiency factor.

The power coefficient $C_p$ is the product of the mechanical efficiency $\eta_m$, the electrical efficiency $\eta_e$, and the aerodynamic efficiency. All three factors are dependent on the wind speed and the produced power. It has been shown by Betz law that the maximum possible value of the aerodynamic efficiency is 0.59, which is achieved when the turbine reduces the wind speed to one-third of the free wind speed (see Chapter 3). This means that a maximum of 59% of the wind energy can be converted into electrical energy by a wind turbine.

The air density can be calculated from air temperature and pressure; the mean air density at a site can be estimated from Fig. 2.1. Power curves are sometimes referenced to 'standard conditions' corresponding to sea level pressure and a mean air temperature of 15°C.

## 2a1. Calculation of annual energy production (AEP)

For a particular wind turbine, power production is specified by the power curve as described in Section 3. By combining the power curve with the wind speed distribution at hub height, the mean annual energy production AEP is given by:

*Fig. 2.1: Air density [kg m³] as a function of altitude/elevation [m] and mean air temperature [°C] at the same altitude / elevation. A lapse rate of 6.5 K/km and a sea level pressure of 1013.25 hPa are assumed.*

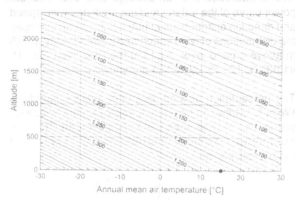

$$\text{AEP} = N_0 \int_{V_{stop}}^{V_{start}} P(u) f(u) du$$

where *P(u)* is the power curve function, *f(u)* the wind speed probability density function, $V_{start}$ the cut-in wind speed, $V_{stop}$ is the cut-out wind speed, and $N_0$ the number of hours in a year. Some terms and definitions are shown in Table 2.1.

*Table 2.1: Wind turbine power production terms and definitions.*

| Term | Definition |
|---|---|
| Full load hours | Number of hours to produce AEP with the turbine running at full power |
| Specific power performance | Annual energy production divided by the rotor area for a specific turbine |
| Capacity factor | Ratio of actual average power to the rated power over the same period of time |
| Potential annual energy output | Assuming distribution of wind speed probability density and 100% availability |

## 2b. Wind speed distribution characteristics

In practice, the power output of a wind turbine or wind farm is estimated over an average year, based on the power curve of the turbine and the predicted wind speed and direction distributions at hub height. As an example, the distributions of wind speed and power density at 125 m above ground level (a.g.l.) at the Risø mast (55° 41' 41.33" N, 12° 05' 22.12" E) are shown in Fig. 2.2.

The presentation of wind data often makes use of the Weibull distribution as a tool to represent the frequency distribution of wind speed in a compact form. The two-parameter Weibull distribution is expressed mathematically as:

$$f(u) = \frac{k}{A}\left(\frac{u}{A}\right)^{k-1} \exp\left(-\left(\frac{u}{A}\right)^{k}\right)$$

where $f(u)$ is the frequency of occurrence of wind speed $u$. The two Weibull parameters thus defined are usually referred to as the scale parameter $A$ and the shape parameter $k$. The influence on the shape of $f(u)$ for different values of the shape parameter is illustrated in Fig. 2.3. For $k > 1$ the maximum (modal value) lies at values $u > 0$, while the function decreases monotonically for $0 < k \leq 1$.

The cumulative Weibull distribution $F(u)$ gives the probability of the wind speed exceeding the value $u$ and is given by the simple expression:

$$F(u) = \exp\left(-\left(\frac{u}{A}\right)^{k}\right)$$

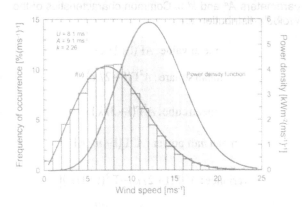

*Fig. 2.2: Distribution of wind speeds and the power density function at 125 m above ground level at the Risø mast for the period 1961-1990. Histograms are observed data; full lines are Weibull fits to the data.*

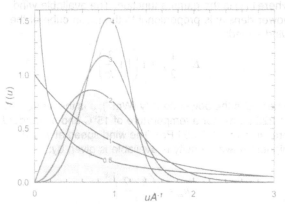

*Fig. 2.3: The shape of the Weibull distribution function for different values of the shape parameter: k = 0.5, 1, 2, 3, 4. Two special cases are: k = 1, the exponential distribution, and k = 2, the Rayleigh distribution.*

The Weibull distribution generates Weibull-distributed higher powers: if $u$ is Weibull-distributed with parameters $A$ and $k$, then $u^m$ is Weibull-distributed with the parameters $A^m$ and $k/m$. Common characteristics of the Weibull distribution are:

$$\text{mean value}: A\Gamma(1+1/k)$$

$$\text{mean square}: A^2\Gamma(1+2/k)$$

$$\text{mean cube}: A^3\Gamma(1+3/k)$$

$$\text{mean } m\text{th power}: A^m\Gamma(1+m/k)$$

$$\text{variance}: A^2\left[\Gamma(1+2/k)-\Gamma^2(1+1/k)\right]$$

$$\text{modal value}: A\left(\frac{k-1}{k}\right)^{1/k}$$

$$\text{median}: A(\ln 2)^{1/k}$$

where $\Gamma(x)$ is the gamma function. The available wind power density is proportional to the mean cube of the wind speed:

$$E = \frac{1}{2}\rho A^3\Gamma\left(1+\frac{3}{k}\right)$$

where $E$ is the power density (Wm$^{-2}$), $\rho$ is the air density (1.225 kg m$^{-3}$ for a temperature of 15°C and a standard pressure of 1013.25 hPa). The wind speed at which the highest power density is available is given by:

$$u_m = A\left(\frac{k+2}{k}\right)^{1/k}$$

Thus, for a Rayleigh distribution, the wind speed which contains the highest energy on the average is twice the most frequent speed (modal value).

## 2c. Global distribution of wind energy

An overview of the global wind climate is illustrated in Fig. 2.4 , where the mean wind speed at 10 m a.g.l. is shown. The picture is a familiar one displaying clearly the "roaring forties" on the Southern Hemisphere and the extra-tropical cyclonic activity over the Northern Atlantic and the Northern Pacific (Petersen, Mortensen, Landberg, Højstrup and Frank (1998; 1998a). Furthermore, the southwest monsoon can be seen, with the Somali Jet standing out, as well as the low-wind regions of the Tropics.

Evidently, this is a very coarse picture of the wind regimes of the World: it does not display local wind systems on scales less than a few hundred kilometers and larger-scale systems with strong yearly variations are suppressed, too. Furthermore, it does not display the pronounced influence of terrain features on scales of less than several hundred kilometers. However, as a starting point for regional wind resource estimation world wide, the database used for the map is useful in combination with adequate meteorological models, see Section 2.e.

## 2d. Wind flow over different terrains

The wind speed at a meteorological station or wind turbine site is determined mainly by two factors: the overall weather systems and the nearby topography out to a few tens of kilometers from the site. In the European Wind Atlas a comprehensive set of models for the

*Fig. 2.4: Mean wind speed in ms⁻¹ at 10 m above ground level for the period 1976-95, according to the NCEP/NCAR reanalysis data set map compiled by H. Frank (Petersen, Mortensen, Landberg, Højstrup, and Frank, 1998).*

for the horizontal and vertical extrapolation of meteorological data and the estimation of wind resources were developed (Fig 2.5); these are implemented in the WAsP PC program (Wind Atlas Analysis and Application Program).

The models are based on the physical principles of flows in the atmospheric boundary layer and they take into account the effect of different surface conditions, sheltering effects due to buildings and other obstacles, and the modification of the wind imposed by the specific variations of the height of ground around the site in question. The effects on the wind speed of these topographic characteristics are described below.

Fig. 2.5: The wind atlas methodology of WAsP. Meteorological models are used to calculate the regional wind climatology from the observed data (the analysis part). In the reverse process (the application of wind atlas data) the wind climate at any specific site may be calculated from the regional climatology (Landberg, Mortensen, Dellwik, et al., 2006; Troen and Petersen, 1989).

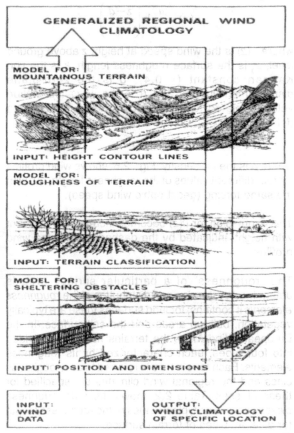

GENERALIZED REGIONAL WIND CLIMATOLOGY

MODEL FOR: MOUNTAINOUS TERRAIN

INPUT: HEIGHT CONTOUR LINES

MODEL FOR: ROUGHNESS OF TERRAIN

INPUT: TERRAIN CLASSIFICATION

MODEL FOR: SHELTERING OBSTACLES

INPUT: POSITION AND DIMENSIONS

INPUT: WIND DATA

OUTPUT: WIND CLIMATOLOGY OF SPECIFIC LOCATION

### 2d1. Wind profile over flat uniform terrain – neutral conditions

At high wind speeds the wind profile over flat and reasonably homogeneous terrain is well modeled using the logarithmic law:

$$u(z) = \frac{u_*}{\kappa}\ln\left(\frac{z-d}{z_0}\right)$$

where $u(z)$ is the wind speed at height $z$ above ground level, $z_0$ is the surface roughness length, $k$ is the von Kármán constant ($\approx 0.4$), $d$ is the zero-plane displacement length, and $u_*$ is the so-called friction velocity related to the surface stress $\tau$ through the definition:

$$|\tau| = \rho \cdot u_*^2$$

where $\rho$ is the air density. Fig. 2.6 shows wind profiles for four different values of the roughness length, but with the same forcing (geostrophic wind speed).

This expression for $u(z)$ becomes increasingly deficient with height, indicated by the dashed parts of the wind profiles.

The "roughness" of a particular surface area is determined by the size and distribution of the *roughness elements* it contains; for land surfaces these are typically vegetation, built-up areas and the soil surface. In the European Wind Atlas, the terrains have been divided into four types, each characterized by its roughness elements. Each terrain type is referred to as a roughness class and the regional wind climates are specified for these classes. Table 2.2 shows the four roughness classes as well as typical values of the roughness length for different types of terrain surface.

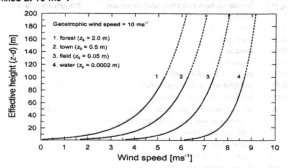

Fig. 2.6: Variation of mean wind speed with height (z) above zero-plane displacement length (d), for neutral conditions and four typical values of surface roughness length, $z_0$. Geostrophic wind speed is fixed at 10 $ms^{-1}$.

## 2d2. Wind profile over flat uniform terrain – non-neutral conditions

Even at moderate wind speeds, deviations from the logarithmic profile occur when $z$ exceeds a few tens of meters. Deviations are caused by the effect of buoyancy forces in the turbulence dynamics; the surface roughness is no longer the only relevant surface characteristic but must be supplemented by surface heat flux parameters. With surface cooling at night time, turbulence is lessened causing the wind profile to increase more rapidly with height; conversely, daytime heating causes increased turbulence and a wind profile more constant with height.

Similarity expressions for these more general profiles are given by:

$$u(z) = \frac{u_*}{\kappa}\left(\ln\left(\frac{z-d}{z_0}\right) - \Psi\left(\frac{z-d}{L}\right)\right)$$

Table 2.2: Typical roughness lengths for different terrain surface characteristics according to the European Wind Atlas (Troen and Petersen, 1989).

| $z_0$ [m] | Terrain surface characteristics (land use) | Roughness Class |
|---|---|---|
| > 1 | tall forest | |
| 1.00 | City | |
| 0.80 | low forest | |
| 0.50 | suburbs | |
| 0.40 | shelter belts | 3 (0.40 m) |
| 0.20 | many trees and/or bushes | |
| 0.10 | farmland with closed appearance | 2 (0.10 m) |
| 0.05 | farmland with open appearance | |
| 0.03 | farmland with very few buildings/trees | 1 (0.03 m) |
| 0.02 | airport areas with some buildings and trees | |
| 0.01 | airport runway areas | |
| 0.008 | mown grass | |
| 0.005 | Bare soil (smooth) | |
| 0.001 | Snow surfaces (smooth) | |
| 0.0003 | Sand surfaces (smooth) | |
| 0.0002 | (used for water surfaces in the Atlas) | 0 (0.0002 m) |
| 0.0001 | water areas (lakes, fjords, open sea) | |

where $\psi$ is an empirical function. The new parameter introduced in this expression is the Monin-Obukhov length $L$:

$$L = -\frac{T_0}{\kappa g} \frac{\rho c_p u_*^3}{H_0}$$

where $T_0$ and $H_0$ are the surface absolute temperature and heat flux, respectively, $c_p$ is the heat capacity of air at constant pressure, $g$ the acceleration of gravity and the remaining quantities are defined above. Fig. 2.7a shows wind profiles for three different stability conditions, where the profiles have been matched at 50 m above ground level, but for the same surface roughness length. Fig. 2.7b, shows new and improved parameterizations of the wind profile for heights between approx. 50 and 200 m a.g.l., compared to measurements at the Danish Test Station for Large Wind Turbines in Høvsøre, Jutland

*Fig. 2.7: Variation of mean wind speed with height above ground level for three different stability conditions: (a) unstable (L = -71 m), neutral (L → ∞) and stable (L = 108 m); (b) normalized (Gryning, Batchvarova, Brümmer, et al., 2007).*

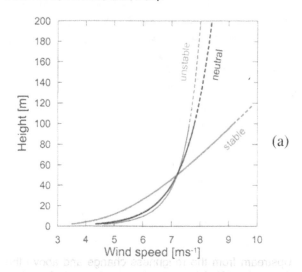

(Gryning, Batchvarova, Brümmer, Jørgensen and Larsen, 2007). The 'classic' wind profiles described above are shown with dashed lines.

### 2d3. Wind profile over flat heterogeneous terrain

If the terrain is not homogeneous and a marked change of terrain surface roughness occurs, an internal boundary layer (IBL) develops downstream from the roughness change. The height of this boundary layer increases with downstream distance (Fig. 2.8).

*Fig. 2.7 b (continued):*

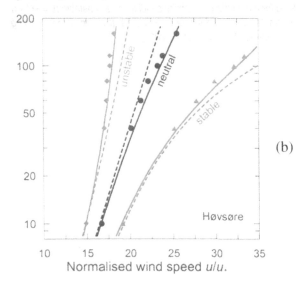

(b)

Normalised wind speed $u/u_*$

Upstream from the roughness change and above the developing IBL (shown by the grey layers in Fig. 2.8) the roughness change is not felt, and the wind speed is determined by the upstream terrain roughness. In the lower part of the IBL (dark grey layer), the wind speed is determined solely by the downstream roughness, and in the light grey layer the wind speed depends on the downstream as well as upstream roughness.

### 2d4. Wind profile over hills and escarpments

If the terrain is not flat and significant changes of terrain elevation occurs, the wind profile is changed according to the elevation and slope of the terrain. An example is

*Fig. 2.8: Change of wind profile following a step change in surface roughness length from $z_0 = 0.0002$ m to $z_0 = 0.10$ m (coastline). Wind is from left to right. Growth of the IBL is indicated by the grey zones (Troen and Peterson, 1989).*

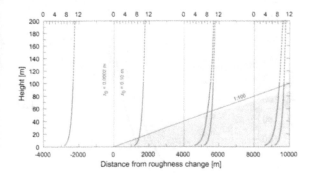

given in Fig. 2.9 (a), where wind profiles upstream, at the hill top and downstream of a simple, 100-m high, bell-shaped (Gaussian) hill have been calculated using the simple flow model of the European Wind Atlas. The height of maximum relative speed-up ($\Delta U/U$) at the hill top occurs below 10 meters, and above this height the speed-up decreases with height. This is also apparent from Fig 2.9 (b) which shows the relative wind speed across the same hill for different heights above ground level. At the hill top, the speed-up vanishes at about $2L$ above the ground, where $L$ is the half-width at the middle of the hill (Troen and Peterson, 1989). It is rarely possible to apply simple rules and formulas to determine the speed-up at specific locations. For this reason it is necessary in most cases to use a numerical model for the calculations.

*Fig. 2.9: Variation of mean wind speed with height above ground level upwind (a), at the hilltop and downwind at the foot of a 100-m high, bell-shaped hill (b) (Troen and Peterson, 1989).*

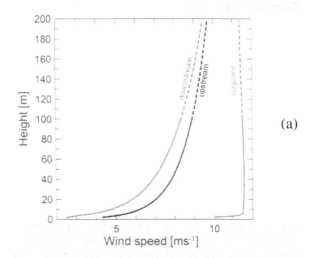

(a)

### 2d5. Wind profile close to sheltering obstacles

Shelter is defined as the relative decrease in wind speed caused by an obstacle in the terrain. Whether an obstacle provides shelter at a specific site depends upon: the distance from the obstacle to the site ($x$), the height of the obstacle ($h$), the height of the point of interest at the site ($H$), the length of the obstacle ($L$) and the porosity of the obstacle ($P$).

Fig. 2-10 shows an example of the reduction of the wind speed due to shelter by a two-dimensional, solid obstacle. The shelter effect vanishes above $3H$ and at a distance of more than $50H$ downstream, where $H$ is the height of the wall.

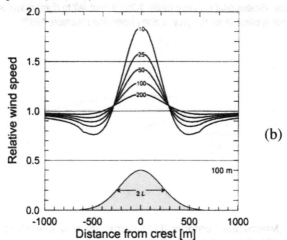

*Fig. 2.9 b (continued):*

**2e. Estimation of wind resources and wind conditions**

Resource assessment and siting of wind farms and turbines are based on both observations and modeling of the wind flow over natural terrain. Fig. 2.11 shows how wind measurements are used together with micro- and meso-scale models to provide reliable predictions of the wind farm annual energy production (AEP) at candidate sites (Hansen, Mortensen, Badger et al., 2007).

Here, different micro- and meso-scale models may be applied through generalized interfaces. Examples of meso-scale models shown in Fig. 2-11 are KAMM (Karlsruhe Atmospheric Mesoscale Model), WRF (Weather Research and Forecasting model), MC2

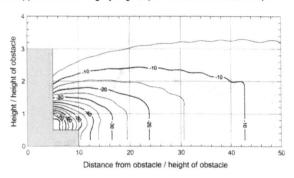

Fig. 2.10: Reduction of wind speed in per cent due to shelter by a two-dimensional obstacle (wall). Wind is from left to right; model is not applicable for the grey region (Troen and Peterson, 1989).

(Mesoscale Compressible Community model) and MM5 (Fifth-Generation NCAR/Penn State Mesoscale Model); sample microscale models shown are WAsP (Wind Atlas Analysis and Application Program), MS-Micro or a CFD model (Computational Fluid Dynamics).

### 2e1. Wind measurements for resource estimation and verification

An overview of instruments for wind speed measurement is presented in Table 2.3.

An important characteristic of the wind energy power is that the power output of a turbine is proportional to the third power of the wind speed. Therefore the precision requirements of wind speed statistics for energy assessments are high. Other requirements for wind data are:

*Fig. 2.11: Wind atlas methodologies use wind measurements as well as microscale and mesoscale modeling for wind resource assessment and siting (Hansen, Mortensen, Badger et al., 2007).*

| | Pre-processing | Modelling | Post-processing | Numerical WA |
|---|---|---|---|---|
| **Mesoscale** | Wind classes | KAMM | Predicted wind climate | Mesoscale maps |
| | Terrain elevation | WRF | Regional wind climate | Database |
| | Terrain roughness | MC2 | Predicted wind resource for selected terrain site coordinates | WAsP *.LIB files |
| | Input specifications | MM5 | | Uncertainties |
| | Model setup | etc. | | Parameters |
| **Measurements** | Met. stations | Wind data | Verification | Applications |
| | Siting | Data collection | Meso- and microscale results vs. measured data | Best practices |
| | Design | Quality control | Adjust model and model parameters to fit data | Courses and training |
| | Construction | Wind database | Satellite imagery (offshore atlas only) | Microscale flow model |
| | Installation | Wind statistics | | Wind farm wake model |
| | Operation | Observed wind climate | | ⇒ Wind farm AEP |
| **Microscale** | Pre-processing | Modelling | Post-processing | Observational WA |
| | Wind speed distribution | WAsP | Regional wind climate | Microscale maps |
| | Wind direction distribution | MS-Micro | Predicted wind climate | Database |
| | Terrain elevation | CFD-models | Predicted wind resource for selected terrain site coordinates | WAsP *.LIB files |
| | Terrain roughness | etc. | | Uncertainties |
| | Sheltering obstacles | | | Parameters |

- The wind data must be *accurate*
  - equipment design and specifications
  - calibration of sensors (anemometers)
  - careful mounting of sensors on mast
- The wind data must be *representative*
  - statistics for several full years (no seasonal bias)
  - data recovery > 90% (missing data at random)
  - careful choice of site for meteorological mast (similarity principle)
- The wind data must be *reliable*
  - verification of sensor outputs, O&M
  - redundant sensors and wind index

## 2e2. Microscale modeling – observational wind atlas

The observed wind climate at a meteorological station can be transformed to a regional wind climate through the wind atlas methodology, as was illustrated in Fig. 2.5. If this is done at a large number stations in a region, an observational wind atlas describing the regional wind climate or energy potential of the region can be obtained. This was done in the preparation of the European Wind Atlas (Fig. 2.12).

The same methodology is used on a smaller scale to obtain resource estimates and wind farm power predictions at candidate wind farm sites.

## 2e3. Mesoscale modeling – numerical wind atlas

If sufficient high-quality data are not available, the regional wind climate may be determined using a mesoscale model. This model determines the wind climate at grid points with a regular spacing of typically 2-5 km. The area covered by one model domain may be on the order of $500 \times 500$ km$^2$. An example is given in Fig. 2.13, which shows the regional wind climate of Egypt determined using the KAMM model (Mortensen, Hansen, Badger, et al., 2005).

This map is based on more than 50,000 estimates of the regional wind climate at the model grid points. Each grid point corresponds conceptually to a virtual meteorological station in the mesoscale model landscape and the regional wind climate is derived by a procedure similar to the analysis of the wind atlas methodology. The modeling results are verified at the Egyptian meteorological stations, where an independent

*Table 2.3: Overview of instruments for wind speed measurements (cup, propeller and sonic anemometers should all be calibrated traceable in an accredited wind tunnel).*

| Instrument | Technique | Measurement | Limitations | Uncertainty* | Cost |
|---|---|---|---|---|---|
| Cup Anemometer (reference IEC61400-12-1) | Drag on cups generates rotation that is measured | Point measurement, measures average horizontal wind speeds | Mast and boom influences, rime on cups around 0°C, bearing wear | Standard deviation 2.5% top-mounted, 3% boom-mounted at 10 m/s | About 600 EURO (large masts expensive, example 100m 250.000 EURO) |
| Propeller anemometer | Lift on propeller generates rotation | Point measurement, measures average horizontal wind speeds | Mast and boom influences, not stable in very turbulent wind, bearing wear | Standard deviation 3% at 10 m/s, boom-mounted and low turbulence | About 1500 EURO |
| SONIC | Acoustic flight times between opposing sensor heads | Point measurement, measures three wind components, turbulence | Mast and boom influences | Standard deviation 3% at 10 m/s boom-mounted | About 2000 EURO for 2D and about 4000 EURO for 3D |
| SODAR | Sound pulses from the ground being reflected from atmosphere at different levels | Larger measurement volume, measures wind shear up to about 150m | No calibration method available, (eventually use mast), influences of surroundings, buildings, turbulence, complex terrain | Standard deviation more than 5% at 10 m/s | About 40.000 EURO |
| LIDAR | Laser from ground scans in a vertical cone and detects Doppler shift from aerosols | Measurement volume in vertical scanning cone, measures wind shear up to about 250m | Calibration method under development (eventually use mast), influences of turbulence and clouds, complex terrain | Expected standard deviation 3% at 10m/s (good correlation with cup anemometer) | About 150.000 EURO |
| Satellite SAR (Synthetic Aperture Radar) | Microwave backscatter from sea surface | Wind maps in the horizontal domain with 2 by 2 km² resolution | Valid at 10 m height, only available over sea. Cover coastal sites. | Standard deviation 1.1 m/s and 20 degrees | 400 EURO per image; around 100 images recommended for pre-feasibility |
| Satellite scatterometer | Microwave backscatter from sea surface | Wind maps in the horizontal domain with 25 by 25 km² resolution | Does not cover coastal sites | Standard deviation 1.5 m/s and 20 degrees | More than 5000 global observations available for pre-feasibility at no charge |

*Under best practice conditions

33

*Fig. 2.12: Observational wind atlas: The European Wind Atlas map depicts the regional wind climate based on observed wind climates at more than 220 stations in 12 EC countries (Troen and Peterson, 1989).*

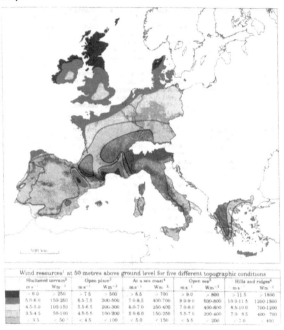

| Wind resources[1] at 50 metres above ground level for five different topographic conditions | | | | | | | | | |
| Sheltered terrain[2] | | Open plain[3] | | At a sea coast[4] | | Open sea[5] | | Hills and ridges[6] | |
| m s⁻¹ | Wm⁻² | m s⁻¹ | Wm⁻² | m s⁻¹ | Wm⁻² | m s⁻¹ | Wm⁻² | m s | Wm⁻² |
| > 6.0 | > 250 | > 7.5 | > 500 | > 8.5 | > 700 | > 9.0 | > 800 | > 11.5 | > 1800 |
| 5.0-6.0 | 150-250 | 6.5-7.5 | 300-500 | 7.0-8.5 | 400-700 | 8.0-9.0 | 600-800 | 10.0-11.5 | 1200-1800 |
| 4.5-5.0 | 100-150 | 5.5-6.5 | 200-300 | 6.0-7.0 | 250-400 | 7.0-8.0 | 400-600 | 8.5-10.0 | 700-1200 |
| 3.5-4.5 | 50-100 | 4.5-5.5 | 100-200 | 5.0-6.0 | 150-250 | 5.5-7.0 | 200-400 | 7.0-8.5 | 400-700 |
| < 3.5 | < 50 | < 4.5 | < 100 | < 5.0 | < 150 | < 5.5 | < 200 | < 7.0 | < 400 |

estimate of the regional wind climate may be obtained. The database underlying the map can be used to reliably estimate the wind climate and wind farm power production anywhere in Egypt, even in the data-sparse desert regions and offshore.

*Fig. 2.13: Numerical wind atlas: The Wind Atlas for Egypt map is based on more than 50,000 regional wind climates in 5- and 7.5-km regular grids, derived by mesoscale modeling. The basic wind climate statistics were obtained from the NCEP/NCAR reanalysis data set for the period 1965-1998. Place names refer to meteorological stations used for verification of the modeling results (Hansen, Mortensen, Badger et al., 2005).*

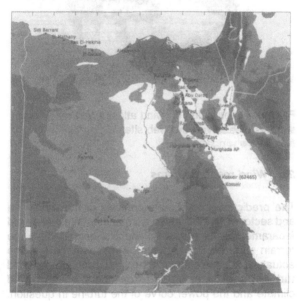

### 2e4. Wind atlases of the world

Measurements for wind energy applications and resource assessments based on modeling have been carried out for several decades in well over 100 countries and territories around the world. Fig. 2.14 summarizes the information available by the end of 2007.

*Fig. 2.14: National wind atlases exist for the 'red' countries and the wind atlas methodology has been applied for regional and local studies in the 'blue' countries. No information was available for the 'light grey' countries by the end of 2007.*

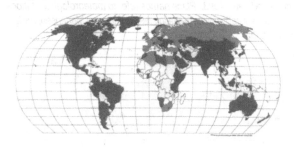

For more information on wind atlases and other wind surveys please visit the web site:
http://www.windatlas.dk.

### 2e5. Wind resource maps

The predicted wind climate consists of the wind rose and sector-wise wind speed distributions (Weibull $A$- and $k$-parameters) for a specific site and height above the terrain surface. The expected wind resource (i.e. the actual exploitable energy in the wind using a given wind turbine) can be estimated from the predicted wind climate and the power curve of the turbine in question. It is usually a simple task to model the variations of the predicted wind climate and resource over a wind farm site. For planning purposes on a regional or national scale, databases and maps of the predicted wind climate over much larger areas are sought. An example of such a database is shown in Fig. 2.15 which shows a wind resource map of Denmark ($\approx$ 43,000 km²).

The Wind Resource Atlas for Denmark 1999 contains wind speed and direction distributions for sites in a regular grid with a horizontal spacing of 200 m (Hansen, Mortensen, Badger et al., 2005). For each point, distributions are given for four heights above ground level: 25, 45, 70 and 100 m. The atlas contains software for data display and retrieval. The accuracy of the Atlas has been verified by comparison with actual wind turbine power productions (Fig. 2.15). The productions for more than 80% of over 1,200 wind turbines are predicted to within ±10%.

## 2f. Calculation and verification of wind resources at specific sites

Wind data analysis, wind resource assessment and siting of wind farms and wind turbines are almost exclusively done by skilled staff using advanced computer and simulation tools. The results, however, are only as good as the inputs provided (Nielsen, 2002). Some of these inputs should be established during site visits to the meteorological stations and wind farm site. Fig. 2.16 may serve as a checklist in connection with such site visits.

## 3. WIND TURBINES

### 3a. Definition of wind turbine

This reference book is concerned only with electricity producing wind turbines and not with small turbines for heating or pumping. Stand alone systems creating local electricity networks are taken into account in some detail.

*Figure 2.15. Wind resource map for Denmark and verification using existing wind turbines (Hansen, Mortensen, Badger et al., 2005; Mortensen and Nielsen, 1999).*

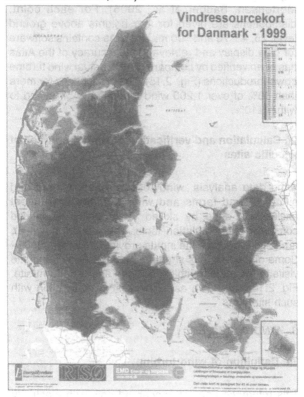

## 3b. Wind turbine concepts

Two main turbine structures are described: Horizontal axis wind turbine (HAWT) and vertical axis wind turbine

*Fig. 2.16: Checklist for inspection of a meteorological station or wind turbine site.*

| Contents of toolbox | Station location |
|---|---|
| ❑ Compass | ❑ Map characteristics |
| ❑ Clinometer | ❑ Geographical coordinates |
| ❑ GPS + spare batteries | ❑ Grid coordinates |
| ❑ Measuring tape | ❑ Site elevation |
| ❑ Camera + spare batteries | ❑ Convergence |
| ❑ Binoculars | **Station description** |
| ❑ Multi-purpose tool, tape | ❑ Photos of station/instruments |
| ❑ WAsP forms and checklist | ❑ Sector photographs (12×30°) |
| ❑ Pocket calculator | ❑ Mast characteristics |
| ❑ Wristwatch or other | ❑ Boom characteristics |
| ❑ Notebook | ❑ Top pole characteristics |
| **Before departure** | ❑ Sensor heights a.g.l. |
| ❑ Topographical maps | ❑ Sensor type, make, serial no. |
| ❑ Site magnetic declination | ❑ Channel allocation checked |
| ❑ Check GPS datum setting | ❑ Cabling and plugs checked |
| ❑ Set wristwatch + PC clock | ❑ System voltage measured |
| ❑ Summaries of wind statistics | ❑ Exchange of data storage |
| ❑ Station manual checked | ❑ Exchange of batteries |
| ❑ Installation report checked | **Before you leave** |
| ❑ Key to station enclosure | ❑ Calibrated data vs. |
| **Upon arrival** |     weather observations |
| ❑ Weather observations | ❑ Toolbox contents complete |
| ❑ Data Acquisition System check | ❑ Data Storing Unit packed |
| ❑ General inspection of station | ❑ Station enclosure locked |

(VAWT) (Fig. 3.1). The HAWT dominates the world market. A HAWT and a VAWT are sketched in Fig. 3.1. The basic components of a HAWT are shown in Fig. 3.2.

*Fig. 3.1: The basic components of the two main types of turbines, HAWT and VAWT. The vertical axis machines include Darrieus, Savonius and Voight-Schneider wind turbines.*

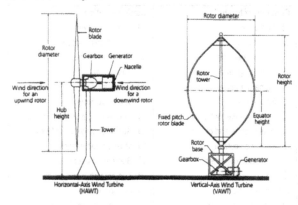

*Fig. 3.2: The basic components of a HAWT with coaxial gearbox. For large turbines, e.g. offshore turbines, a high voltage transformer may be added as a component inside the nacelle.*

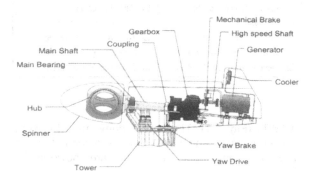

### 3b1. Horizontal axis turbines

Characteristics of horizontal axis turbines: Number of rotors, blade control, electrical and transmission systems:

- One to four rotor blades. HAWTs with three blades turning clock-wise (seen from the wind direction) are dominating the market.
- Control of rotor by *stall-controlled blades* or *variable-pitch blades*. Stall-controlled rotors are supplemented by movable *tip brakes*.
- *Yaw systems* orient the plane of the rotor perpendicular to the wind stream.
- Electrical system may be based on *induction generators* or *dc-generators with converters*.
- *Transmission system* may include *gears* in connection with induction generators or *direct drive operation* in connection with multipole induction or synchronous generators and conversion systems (variable speed turbines).

Development of HAWT rated power capacity: The historical development of HAWT rated power capacity in Denmark is shown in Fig. 3.3.

Higher capacities in the range from 5 to 10 MW are foreseen before 2010 with special application in offshore wind farms. The EU has launched a large project UPWIND aiming to improve existing technology and increasing the sizes of wind turbines up to 20 MW.

The integrated gate bipolar transistor (IGBT) has replaced the thyristor as the dominating inverter component. The thyristor is a semi-conductor device acting as a switch when the gate receives a current pulse

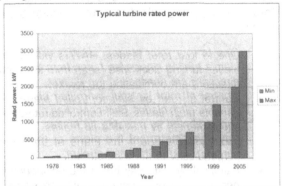

continuing as long as the voltage across the device is not reversed. The IGBT is a fast acting transistor being able to work at very high voltages.

The development in turbine size is illustrated in Fig. 3.4.

### 3c. Direct drive technologies

An interesting concept competing with the dominate HAWT with a gearbox in the transmission system, is the direct drive wind turbine. Table 3.1 shows examples of wind turbines with traditional gearboxes and an induction generator (Vestas and Repower), a direct drive wind turbine (Enercon and Multibrid) wind turbine with one step-up in the gearbox and a multiple generator.

As can be seen, the direct drive wind turbine has the largest mass but the Multibrid 5000 wind turbine which has only a one-step gearbox can compete with the

42

*Fig. 3.4: The development in wind turbine sizes from 1985 and a forecast about future sizes. In only a few years a rotor size of 80 meters is expected to be approached. This size of turbine is shown in the figure in comparison with the largest present day air plane (European Commission, 2006).*

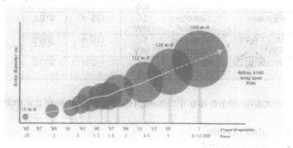

traditional wind turbines with a conventional gear/ generator design.

### 3d. Stall and pitch control of wind turbines

Wind turbines are controlled to limit the power output and the loading on the turbines. The power control is divided into two regimes with different concepts:

- power optimization for low wind speeds
- power limitation for high wind speeds

These regimes are separated by the wind speed at which the maximum power output is achieved, typically from 12 to about 15 m/sec.

Basically, there are three approaches to power control:

*Table 3.1: Mass of turbine top (nacelle and rotor) for different wind turbine concepts.*

| Manufacturer | Concept | Rated power | Rotor diameter | Top mass |
|---|---|---|---|---|
| Enercon | Gearless | 4.5 MW | 112 m | 500 t |
| REpower | Gear/generator | 5.0 MW | 125 m | 410 t |
| MULTIBRID 5000 | One step gearbox | 5.0 MW | 116 m | 300 t |
| Vestas V120 | Gear/generator | 4.5 MW | 112 m | 210 t |

Basically, there are three approaches to power control:

- stall control
- pitch control
- active stall control

Stall-controlled wind turbines have their rotor blades bolted to the hub at a fixed angle. Some drawbacks of this method are: lower efficiency at low wind speeds; no assisted start; and, variations in the maximum steady state power due to variation in the air density and grid frequencies.

Pitch-controlled wind turbines have blades that can be pitched to an angle where the blade chord is parallel to the wind direction. The power output is continuously monitored, and when the wind speed is above a rated value, the control limits the power output. Nowadays, pitch control of wind turbines is only used in conjunction with variable rotor speed. An advantage of this type of control is that it has an efficient power control, i.e. that the mean value of the power output is kept close to the rated power of the generator at high wind speeds. Disadvantages encompass extra complexity due to the pitch mechanism and high power fluctuations at high wind speeds.

Active stall-controlled turbines resemble pitch-controlled turbines by having pitchable blades. At low wind speeds, active stall turbines will operate like pitch-controlled turbines. At high wind speeds, they will pitch the blades in the opposite direction of what a pitch-controlled turbine would do and force the blades into stall. This control type has the advantage of having the ability to compensate for the variations in the air density.

Fig. 3.5 shows iso-power curves for a wind turbine as a function of the blade angle and the mean wind speed. The ranges for pitch control and active stall control are separated at a blade angle of 0° with the rotor plane. The dashed curves illustrate how the transition between operation with 0° blade angle at low wind speeds and power-limiting operation along an iso-power curve at high wind speeds can be achieved for a three-bladed rotor at a rated power of 400 kW. In this example, the rated power is reached at a wind speed of about 12 m/ sec.

Fig. 3.5 illustrates that only a small shift in blade angle is needed to regulate the power output for an active stall controlled turbine compared to that of a pitch-controlled turbine. Therefore the active stall controlled turbine is able to counteract power peaks very efficiently without changing its rotational speed whereas there is a greater need to construct a pitch-controlled turbine as a variable speed machine.

*Fig. 3.5: Iso-power curves for a wind turbine at 26.88 rpm vs. blade angle and mean wind speed*

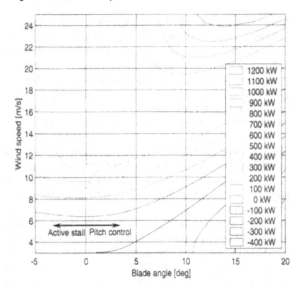

### 3e. Variable speed wind turbines

Variable speed wind turbines have been introduced to reduce loads in the transmissions system (gear loads) and loads on the blades. A variation in rotational speed of some 10–20% means that mechanical stresses (moments) are reduced dramatically. At the same time the overall efficiency of the turbine increases. Modern regulation technology also gives rise to reduced stresses during operation by providing possibilities for fast changes in the shaping of the rear edge.

*Fig. 3.6. Power and efficiency curve for a typical well-sited turbine.*

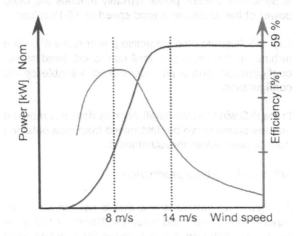

## 3f. Wind turbine performance: Calculations, measurements and wind turbine energy production calculation

### 3f1. Power curve

A power curve gives the relation between the wind speed and the power output of a wind turbine. Fig. 3.6 shows a power curve and the efficiency of a wind turbine. The efficiency is the relation between the power output of a wind turbine and the power in the wind in the rotor swept area at a certain wind speed. The actual value of the power depends on the size of the turbine

The efficiency factor $C_p$ typically reaches a maximum at a wind speed of 7-9 m/sec and normally does not exceed 50%. The theoretical maximum aerodynamic efficiency (not taking mechanical and electrical loads into account) is 59%. The electric power typically reaches the rated power of the turbine at a wind speed of 12-15 m/sec.

Fig. 3.7 illustrates the controlled power curve of a wind turbine, in the case of a stall controlled, fixed speed configuration, and a pitch-controlled, variable speed configuration.

In the IEC wind turbine certification system, it is required that the power curve be determined from data obtained from turbine power measurements.

### 3f2. Single turbine production

A single wind turbine is one that is not affected from wakes from other wind turbines. Normally the wind turbine capacity on a site is given for a single wind turbine. The power production onshore and offshore is illustrated on Fig. 3.8. It can be seen that the energy production decreases rather dramatically as a function of the distance from the coast.

### 3f3. Wind farm

Normally large wind farms are built in low population areas on land or offshore. In the USA large wind farms are built on the Great Plains (in the country's Midwest) and mountain areas. Similar site principles are used in Spain.

### 3f31. Onshore and offshore wind farms

Accurately modeling of wakes is important to optimize wind farm layouts and to evaluate optimal control

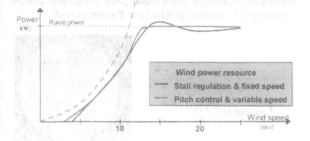

*Fig. 3.7: Examples of power curves for two types of wind turbines.*

strategies. Further development and verification of computational wake models are needed.

### 3f32. Wind farm clusters

Clusters of 4-6 large wind farms placed at short distances from each other will optimize grid costs. Information indicating power losses in wind farms and wind clusters, including measurements, data analysis and modeling to predict shadow effects of large wind farms is given in Fransden, Barthelmie, Rathmann et al. (2007).

### 3g. Aerodynamic, structural and electrical design and verification design tools (dynamics, measurements and verification)

A combination of calculation and testing are used to demonstrate that the structural elements of a wind turbine meet the prescribed level of safety.

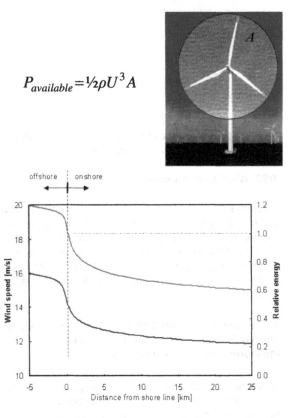

Fig. 3.8: Example of wind speed and energy flux for onshore and offshore single turbines. The available power (P) is found from the air density $(\rho)$, the wind speed (U) and the rotor area (A).

$$P_{available} = \frac{1}{2}\rho U^3 A$$

The **blue (lower) line** is wind speed and the red line is energy flux through the swept area of the wind turbine rotor, relative to the flux at open sea. The exact form will depend of the actual roughness class.

### 3h. Wind turbine standards

Since 1987 standardization work has been hosted by International Electrotechnical Committee (IEC) in IEC-TC-88. Both standards for safety and standardization have been published. Table 3.2 illustrates a selection of published standards both by IEC and by individual countries.

### 3i. Type certification of wind turbines and project certification

The IEC has made a standard for both type certification of wind turbines and project certification. Figs. 3.9 and 3.10 illustrate the modules in the type certification and the project certification.

## *4. WIND TURBINES AND ENERGY SYSTEMS*

A number of different wind power systems are described in this chapter.

### 4a. Stand-alone systems

One example of a stand-alone system is shown in Fig. 4.1 including the following technical outline:

- Back-to-back converter for wind turbine and grid control
- Optional battery in the dc-link
- Controller for system and wind turbine supervisory control

*Table 3.2: Wind turbine standards*

| Standard No. | Identification | Authority |
|---|---|---|
| DS472, 1st edition | Load and Safety for Wind Turbine Structures | Dansk Standard, Dansk Ingeniorforening, Copenhagen, Denmark, 1992 |
| IEC61400-1 2nd edition | Wind turbine generator systems – Part 1: Safety requirements | International Electrotechnical Commission, Geneva, Switzerland, 1999 |
| NVN11400-0 1st edition | Wind turbines – Part 0: Criteria for type certification – technical criteria | Nederlands Normalisatie-instituut, The Netherlands, 1999 |
| DIBt RICHTLINIEN | Richtlinie. Windkraftanlagen. Einwirkungen und Standsicherheits- nachweise für Turm und Gründung (Guidelines for loads on wind turbine towers and foundations) | Deutsche Institut für Bautechnik (DIBt), Berlin, Germany, 1993 |
| GL REGULATIONS | Regulation for the Certification of Wind Energy Conversion Systems," Vol. IV – Non-Marine Technology, Part 1 – Wind Energy, in "Germanischer Lloyd Rules and Regulations | Hamburg, Germany, 1993 |

- Prototype controller
  - Turbine control (rotational speed control)
  - Grid control (start up, sync, (f, U))
  - Dynamic unit control (incl. braking resistor and battery)

*Fig. 3.9: Modules of type certification (from WT01 edition 1)*

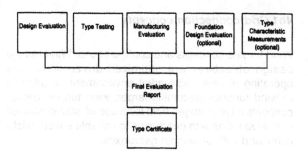

*Fig. 3.10: Modules of Project Certification (from WT01 edition 1).*

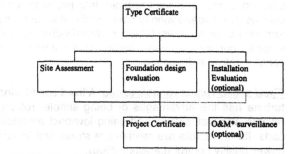

O&M*: Operation and Maintenance

## 4b. Grid-connected wind turbines

### 4b1. Grid connection concepts

Wind turbine concepts and configurations: Wind turbine design objectives are optimized-driven within the operating regime and market environment. In addition to wind turbines becoming larger, wind turbine design concepts have changed from fixed speed, stall-controlled and drive trains with gearboxes to variable speed, pitch-controlled with or without gearboxes.

Wind turbines can be classified according to their *speed control* and *power control* ability. The speed control criterion divides the wind turbines into two significant classes (fixed speed wind turbines; variable speed wind turbines) while the power (blade) control divides the wind turbine concepts in three classes (stall control; pitch control; active stall control).

*Fixed speed versus variable speed:* A fixed speed wind turbine has the advantages of being simple, robust, reliable and well proven and using low-cost electrical parts. Its drawbacks are mechanical stress and limited power quality control (Larsson, 2000).

Variable speed wind turbines have become the dominant type among installed units since 2000 (Hansen and Hansen, 2007). The variable rotational speed implies that the generator has to be decoupled from the grid frequency through power electronics.

By variable speed operation, it is possible to continuously adapt (accelerate or decelerate) the rotational speed of

*Fig. 4.1: Small wind turbine stand-alone system: <u>Main turbine data:</u>*
*11kW (rated power); 2 bladed, diameter ; 13m fixed pitch stall control;*
*Teeter hub; Down wind, passive yaw; Induction generator; Hub*
*height: 18m.*

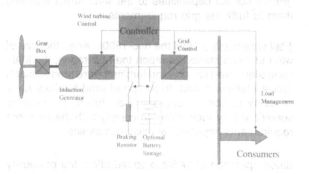

the turbine to the wind speed in such a way that it
operates continuously at its highest level of aerodynamic
efficiency. In contrast to fixed speed wind turbines, which
are designed to obtain maximum efficiency at one wind
speed only (or in some cases two speeds), the variable
speed wind turbines are designed to achieve maximum
aerodynamic efficiency over a wide range of wind
speeds.

Comparing the fixed speed versus variable speed
concepts, the variable speed wind turbine has the
advantages of reduced mechanical stress, increased
power capture, and reduced acoustical noise and high
control capability, which is a prime concern with grid
integration of large wind farms. Owing to these
advantages, variable speed operation is the predominant
choice for MW-scale turbines today. Its direct drawbacks
are additional losses due to the power electronics and
their increased capital cost, and more components.

The main trend of modern wind turbines/wind farms is the variable speed operation and a grid connection through power electronic interfaces. The presence of power electronics inside wind turbines/wind farms offers greater control capabilities to the wind farms enabling them to fulfill the grid requirements.

*Stall versus pitch:* Until the mid-1990s, when the size of wind turbines started to reach the MW range, the stall-controlled, fixed speed wind turbines were predominant. Active stall-controlled, fixed speed wind turbines have recently become popular. Their future success is conditioned by their ability to comply with the stringent requirements imposed by utility companies.

Since operation at variable speed offers the possibility of increased "grid friendliness," there is increasing focus on this concept. The pitch control has proven to be an attractive option for variable speed operation, especially due to power limitation considerations (Hansen, 2004).

The most commonly applied wind turbine designs in the industry can be categorized into four main wind turbine concepts as illustrated in Fig. 4.2 and described below (Hansen and Hansen, 2007).

*Type A - fixed speed wind turbine concept:* This configuration denotes the fixed speed controlled wind turbine, with a Squirrel Cage Induction Generator (SCIG) directly connected to the grid through a transformer. It is known as the "Danish concept," because it was developed and widely used in Danish wind turbines. Since SCIG always draws reactive power from the grid, this concept uses a capacitor bank for reactive power compensation. Smoother grid connection occurs by incorporating a soft-starter.

*Type B - variable speed wind turbine concept with variable rotor resistance:* This configuration corresponds to the limited variable speed controlled wind turbine with Controlled Rotor Resistance Induction Generator (CRRIG) and pitch control. The rotor winding of the generator is connected in series with a controlled resistance, whose size defines the range of the variable speed (typically 0-10% above synchronous speed). Since the losses are proportional to the slip, a limited slip range is used. As for Type A, reactive power compensation and a soft-starter are required for this concept.

*Type C - variable speed wind turbine concept with partial-scale frequency converter (DFIG):* This configuration denotes the variable speed controlled wind turbine with a Double Fed Induction Generator (DFIG) and pitch control. The stator is directly connected to the grid, while the rotor is connected through a partial-scale frequency converter, typically with a capacity of only 25-30% of a full-scale converter.

This concept supports a wider range of dynamic speed control compared to Type B. Typically, the variable speed range comprises +/- 30% around the synchronous speed. The partial-scale frequency converter also provides reactive power compensation and smooth grid connection.

*Type D - variable speed concept with full-scale frequency converter:* This configuration corresponds to the full variable speed, pitch-controlled wind turbine, with the generator connected to the grid through a full-scale frequency converter. The frequency converter provides the reactive power compensation and a smooth grid connection for the entire speed range. The generator

*Fig. 4.2: Typical wind turbine concepts: Types A, B, C and D.*

**Type A**

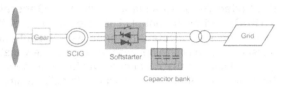

**Type B**

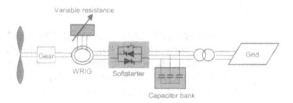

**Type C**

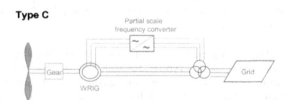

**Type D**

can be electrically excited Wound Rotor Synchronous Generator (WRSG) or Wound Rotor Induction Generator (WRIG) or permanent magnet excited Permanent Magnet Synchronous Generator (PMSG). Typically, a direct driven, multi-pole synchronous generator (no gearbox) is used. This concept has full control of the speed range from 0-100% of the synchronous speed. However, it has a higher cost and a higher power loss in the power electronics, since all the generated power has to pass through the frequency converter.

Over the years, the interest for wind turbine concept A and B has decreased slightly in favor of variable speed wind turbine concepts (Type C and Type D), which today dominate the wind power market. The different control concepts are listed in Table 4.1 while abbreviations of electrical terms are given in Table 4.2.

### 4b2. Power quality (frequency, voltage, flicker)

In electrical networks variations of frequency and voltage is part of the control mechanisms to maintain the power balance and to control and regulate the amount of reactive power in the system.

Grid-connected wind turbines do affect the power quality of the grid. The term power quality of a wind turbine describes the electrical performance of the turbine's electricity generating system in its interaction with the grid. A perfect power quality means that the voltage and current are continuous and sinusoidal with a constant amplitude and frequency. Low reactive power flow is also often desired.

*Table 4.1: Control concepts for wind turbines*

| Power Converter | Mutipole or gearbox | Power control Features | Comments |
|---|---|---|---|
| Battery bank | Gearbox | Stall or active stall | One or two speed machine |
| Frequency converter | Gearbox | Stall or active stall | Variable speed |
| Power Electronic (PE) converter or passive components | Gearbox | Pitch | Limited variable speed |
| Frequency converter | Gearbox | Pitch | Variable speed (doubly fed generator) |
| Frequency converter | Multipole | Stall, active stall or pitch | Variable speed |
| Rectifier and frequency converter | Multipole | Pitch | Variable speed without gearbox |

Power quality depends on the grid and wind turbines characteristics, i.e. grid type (weak, strong), wind turbine technology and the size of wind farms. The increase of wind energy penetration into the power system over the years has influenced the focus on the power quality issues (Sørensen, Cutululis, Lund et al., 2007) which can be divided into:

- *local issues* related to the voltage quality in the distribution systems
- *global issues* related to the power system control and stability.

With international trade of wind power, methods to measure and quantify power quality of wind turbines require international standards. The international

*Table 4.2: Frequently used abbreviations for electrical terms.*

| | |
|---|---|
| AG | Asynchronous Generator |
| AHF | Active Harmonic Filters |
| ASVC | Advanced |
| CUPS | Custom Power Systems |
| CSC | Current Source Converter |
| CSI | Current Source Inverter |
| | Current Stiff Inverter |
| DF | Doubly Fed |
| DPM | Discrete Pulse Modulation |
| DVR | Dynamic Voltage Restorer |
| ETM | Grid Model |
| FACTS | Flexible AC-systems |
| GTO | Gate Turn Off transistor |
| HVDC | High Voltage Direct Current |
| IGBT | Insulated Gate Bipolar Transistor |
| IGCT | Insulated Gate Commutated Thyristors |
| IM | Induction Machine |
| MCT | Mos Controlled Thyristors |
| NCC | Natural Clamped Converter |
| PCC | Point of Common Coupling |
| PE | Power Electronic |
| PMG | Permanent Magnet Generator |
| PSD | Power Spectral Density |
| PWM | Pulse Width Modulation |
| RMS | Grid Model |
| SCIG | Squirrel Cage Induction Generator |
| SG | Synchronous Generator |
| SRG | Switched Reluctance Generator |
| SRM | Switched Reluctance Machine |
| STATCOM | Static Synchronous Compensator |
| SVC | Static Var Compensator |
| UPQC | Unified Power Quality Conditioner |
| UVC | Unified Voltage Controller |
| VSI | Voltage Source Inverter |
| | Voltage Stiff Inverter |
| WRSG | Wound Rotor Synchronous Generator |

standard for power quality measurement of wind turbine IEC 61400-21, 2002 defines such power quality characteristics and proposes corresponding methods to asses the impact of one or more wind turbines on the grid power quality.

There are three possible grid interferences caused by individual or small cluster wind turbines:

- voltage distortions
- frequency distortions
- failures

Voltage distortions within the grid pose the main interference with local power quality generated by wind turbines. This is described in different time intervals as follows:

- _steady-state voltage variations_ are changes in the RMS value of the voltage occurring with a frequency less than 0.01 Hz.
- _flicker_ are voltage fluctuations between 0.01-35 Hz that can cause visible variations in domestic lighting.
- _harmonics_ are voltage fluctuations above 50 Hz, caused by the presence of power electronics (frequency converter).
- _transients_ are random voltage fluctuations caused by e.g. connection of capacitor bank.

The frequency of large power systems is usually very stable, and therefore the individual or small clusters of wind turbines do not typically influence the frequency. However, this is not the case for small autonomous grids, where the wind turbines may cause frequency variations. Wind turbines do not normally cause any interruptions or failures on a high voltage grid.

Depending on the grid configuration and wind turbine concept, different power quality problems may arise. For example, if the wind turbine is fixed speed, the natural variations of the wind and the tower shadow will result in fluctuating power, which may cause flicker disturbances. This effect is usually only relevant for fixed speed wind turbines, while it is absorbed in variable wind turbines as kinetic energy. In IEC 61400-21, only variable speed wind turbines must be tested for harmonics, due to the presence of the frequency converter, which may inject harmonic currents into the grid.

### 4c. Integration of fluctuating wind power in the electrical grid

#### 4c1. Power control of grid

Modern wind turbine technology makes it possible to control both active and reactive power within a limited power value.

The general structure of a simulation model of a wind farm controller with remote access is shown on Fig. 4.3. Requests from the power system operator or the wind farm owner are specified as input to the controllers. A wind farm controller measures the characteristics of the power generated by the wind farm at the point where it connects to the grid. It then controls this power by issuing instructions to the individual wind turbines, each of which has its own controller. The wind farm controller also receives information from the turbines about the maximum amount of power available.

*Fig. 4.3: The general structure of a wind farm control system. Level of system operator. PCC is the Point of Common Coupling.*

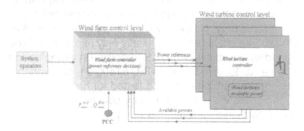

Models are developed for many different wind turbine technologies, taking into account different electrical design and control strategies.

In Fig. 4.4 the power output from an entire wind farm is compared to that from a single turbine, measured over a two-hour period. Summing up the outputs from many wind turbines effectively smooth out fluctuations in the short term. However, this is only valid for short term fluctuations meaning that at long timescales the large scale fluctuations persist.

## 5. OFFSHORE WIND POWER

The implementation of offshore wind farms is now spreading to many waters in Europe as illustrated in Table 5.1.

The size of the turbines for offshore farms has been increasing with time in order to reduce the relative cost of the foundation and substructure (the substructure is the structure between the tower and the foundation).

*Fig. 4.4: Power fluctuations from a single wind turbine compared to the entire wind farm. The coherence function is the key to quantifying the geographical smoothing effect, which can explain why a 160 MW wind farm can contribute more power fluctuations than 2400 MW of wind turbines distributed over a much larger area. Accurate quantification of the large-scale coherence effect is therefore essential to reliable models.*

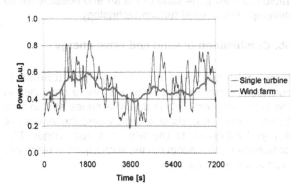

of the foundation and substructure (the substructure is the structure between the tower and the foundation).

## 5a. Economics of offshore wind farms

The advantage of offshore farms compared to land-based wind farms is a higher and more stable wind regime. This may give rise to about 50% higher annual production offshore than for land-based turbines of the same rated capacity. A comparison of investment costs is given in Table 5.2 for two different cases of offshore turbines, in both shallow waters (15 m depth) and deep water position (30 m depth).

The disadvantages of offshore farms are higher investment and maintenance costs. In 2005 a tender for a Danish 200 MW offshore wind farm at Horns Reef in the North Sea resulted in a tariff of about 7 Eurocents/kWh as compared to a production cost at good land-based sites of about 4.5 Eurocents/kWh.

Table 5.3 shows the state of the art and possible future development in wind turbine technology.

## 5b. Combination of wind and wave energy

There may be some interesting opportunities in combining offshore wind farms with wave machines at the same location. This is due to a phase shift in time between the two kinds of energy, where the waves are delayed relatively to the wind but last longer. The combination will smooth out the fluctuations in electricity production to some extent.

## 6. ECONOMICS OF WIND POWER

The size of the generator of a wind turbine plays a fairly minor role in the pricing of a wind turbine, even though the rated power of the generator tends to be approximately proportional to the swept rotor area. Some important economic and physical relations are shown in Tables 6.1 and 6.2.

*Table 5.1: Offshore wind farms globally at the end of 2006 (BTM Consult ApS, 2007).*

| Year | Place | Rated capacity | Number of turbines | Total capacity MW | Supplier | Foundation Type |
|------|-------|----------------|--------------------|-------------------|----------|-----------------|
| 1991 | Vindeby (DK) | 450 kW | 11 | 5 | Bonus | Concrete caisson |
| 1994 | Lely (NL) | 500 kW | 4 | 2 | Micon | Driven monopole |
| 1995 | Tunø Knob (DK) | 500 kW | 10 | 5 | Vestas | Concrete caisson |
| 1996 | Drotten Isselmeer (NL) | 600 kW | 28 | 17 | Micon | Driven monopole |
| 1997 | Bockstigen (S) | 550 kW | 5 | 3 | Micon | Drilled monopole |
| 2000 | Utgrunden (S) | 1.5 MW | 7 | 10.5 | GE Wind | Driven monopole |
| 2000 | Blyth (UK) | 2 MW | 2 | 4 | Vestas | Drilled monopole |
| 2000 | Middelgrunden DK) | 2 MW | 20 | 40 | Bonus | Concrete caisson |
| 2001 | Yttre Stengrund (S) | 2 MW | 10 | 10 | Micon | Drilled monopole |
| 2002 | Horns Reef (DK) | 2 MW | 80 | 160 | Vestas | Driven monopole |
| 2002 | Paludan Flak (DK) | 2.3 MW | 10 | 23 | Bonus | Driven monopole |
| 2003 | Nysted Havmøllepark (DK) | 2.3 MW | 72 | 166 | Bonus | Concrete caisson |
| 2003 | Arklow Bank Phase I (IRL) | 3.6 MW | 7 | 25 | GE Wind | Driven monopole |
| 2003 | North Hoyle (UK) | 2 MW | 30 | 60 | Vestas | Driven monopole |
| 2004 | Scroby Sands (UK) | 2 MW | 30 | 60 | Vestas | Driven monopole |
| 2005 | Kentish Flat (UK) | 3 MW | 30 | 90 | Vestas | Monopile |
| 2006 | Barrow (UK) | 3 MW | 30 | 90 | Vestas | Monopile |
| 2006 | NSW (NL) | 3 MW | 36 | 118 | Vestas | Monopile |

*Table 5.2: Distribution of investment costs for turbines onshore and offshore for two different water depths at typical shallow water (15 m) and deep water (30 m).*

| Turbine and external costs | Share of total cost, % | | |
|----------------------------|---------|---------|------|
| | Onshore | Offshore | |
| | | Shallow | Deep |
| Turbine (ex works) | 78 | 40 | 31 |
| Foundation | 4 | 20 | 36 |
| Electric, incl. grid connection | 10 | 28 | 22 |
| Other | 8 | 12 | 11 |

*Table 5.3: State of the art and possible future development in wind turbine technology. H is hub height, D is rotor diameter and $P_{max}$ is maximum power output from unit.*

| Issue | State of the Art | Future Trend |
|---|---|---|
| Size | $H$=100m, $D$=100m, $P_{max}$=3MW | $H$=150m, $D$=150-200m, $P_{max}$=10MW |
| Operation & maintenance | Costs over life time correspond to approx. 30-50% of initial investment | Halving of O&M cost |
| Water depth | 5-20m | 10-50m |
| Foundation/ Substructure | Mono-pile and gravity founding; bucket; tripod | (Far future: various floating concepts) |
| Blade tip speed | 60-80 m/s | 80-120 m/s |
| Structural design | Passive; some active control | Active control: limitation of loads on all structural components |
| Materials | Chosen for strength | In addition also as consequence of lifecycle analysis |
| Control | Separate wind turbine/farm | In conjunction with regional grid; commercial optimization |
| Grid | Normal terminal; national grid | Island-operation; storage; international connections |
| Production forecast | Market requirement is 12-36 hours | Market requirement reduced to 2 hours |
| Production strategy | Maximum energy | Maximum revenue |
| Transmission | AC – alternate current | HVDC – high voltage direct current |

*Table 6.1: Useful Rules of Thumb for Wind Turbine Economic Relations*

| Wind Turbine price | |
|---|---|
| Proportional to | Square of blade length |
| Proportional to | Swept rotor area |
| Proportional to | Third root of hub height |

*Table 6.2: Useful Rules of Thumb for Energy Production. Tip Speed Ratio is defined as the ratio between speed of wing tip and the wind speed.*

| Wind Turbine Energy Production and Tip Speed | |
|---|---|
| Proportional to | Rotor area at given tip speed |
| Proportional to (roughly) | Square root of hub height |
| Tip speed ratio limit | 75 m/s for wind noise reasons |

The choice of rotor size and generator size depends on the distribution of the wind speed and the wind energy potential at a prospective location. A large rotor fitted with a small generator will produce electricity during many hours of the year, but it will only capture a small part of the wind energy potential. A large generator is very efficient at high wind speeds, but inefficient at low wind speeds. Sometimes it is beneficial to fit a wind turbine with two generators with different rated powers as this make it possible to use a wider spectrum of wind velocities.

## 6a. Investment

Capital costs of wind energy projects are dominated by the price of the wind turbine itself. In Table 6.3 the cost structure is shown for components of the turbine, and Table 6.4 shows turbine versus external costs for a medium sized turbine sited on land. The average turbine share of the total project cost is typically a little less than 80% for an onshore sited turbine but with some variation ranging from about 74% to about 82%.

The costs of wind turbine installation include notably:

- Foundations
- Road construction

*Table 6.3: Relative costs of main parts in a typical wind turbine (Specific reference values of 45.3 meter blade length and 100 meter tower). A typical wind turbine will contain up to 8,000 different components (Wind Directions, 2007).*

| Turbine component cost | Average share of total cost % |
|---|---|
| Tower | 26 |
| Rotor blades | 22 |
| Gearbox | 13 |
| Power converter | 5 |
| Transformer | 4 |
| Generator | 3 |
| Main frame | 3 |
| Pitch system | 3 |
| Main shaft | 2 |
| Rotor hub | 1 |
| Brake system | 1 |
| Yaw system | 1 |
| Rotor bearings | 1 |
| Nacelle housing | 1 |
| Other minor parts | 14 |

*Table 6.4: Cost structure for a typically medium size wind turbine (850 kW – 1500 kW).*

| Turbine and external costs | Share of total cost, onshore (Average) % |
|---|---|
| Turbine (ex works) | 78 |
| Foundation | 4 |
| Electric installation | 4.5 |
| Grid-connection | 5,5 |
| Consultancy | 1.5 |
| Land | 1.5 |
| Financial costs | 2.5 |
| Road construction | 2.5 |

*Based on a limited data-selection (UK, Spain, Germany*

- Underground cabling within the wind farm
- Low to medium voltage transformers
- (Possibly) medium to high voltage substation
- Transport, craning
- Assembly and test
- Administrative, financing and legal costs

Generally speaking, there are _economies of scale_ in the construction of wind farms, both in terms of the _total size of the wind farms_ (the number of turbines sharing a common substation and sharing development and construction costs) – and in terms of the size of turbines. Larger turbines generally have comparatively lower installation costs per swept rotor areas, and the cost of a number of wind turbine components such as electronic controllers and foundations _vary less than proportionately with the size of the wind turbine._

## 6b. Economic and technical lifetime

Wind turbines are usually type certified to withstand the vagaries of the particular local wind climate class safely for 20 years, although they may survive longer, particularly in low-turbulence climates.

A number of life cycle analyses have documented a maximum energy pay back time of 8 months for wind turbines. This figure includes all purchase and services requirements expressed equivalently in units of energy during a predicted lifetime of a wind turbine of 20-25 years. This result is not sensitive to whether the turbine is positioned onshore or offshore.

# 7. ECONOMIC SUPPORT SCHEMES AND INCENTIVES

## 7a. Economic support schemes

The main economic support schemes and incentives are listed in Table 7.1 Some variations of the schemes have been observed as shown in Meyer (2003). Also different names appear in different countries.

## 7b. Global distribution of support schemes

Different types of support schemes have been used in Europe and the US in order to promote the penetration of wind power.

*EU strategies:* The European Union (EU) Commission has discussed the possibility of harmonizing the support schemes used by Member States but has postponed such a decision until more experience with the different systems have been obtained. Increasing oil prices and decreasing investment costs for wind power may, however, eliminate the need for any support scheme before a decision about harmonization is on the table.

*USA strategies:* At federal level a scheme named Production Tax Credits (PTC) has been used to promote the penetration of wind power. This scheme has been handicapped by a "stop-and go" character due to short periods (two to three years) between political decisions about the future of the scheme.

Table 7.1: Main economic support schemes and incentives for promotion of wind power.

| Name | Characteristics | Remarks | Global distribution |
|------|-----------------|---------|---------------------|
| Investment subsidy | Part of investment cost. Subsidy depending on certification, approval or positive test by authorities | Does not encourage efficiency and continued operation, but improves technology transfer from public R&D | Not very frequent |
| Tax reductions | Deduction of part of investment cost | Often lack of transparency | Used in the US and partly in Europe |
| Fixed Feed-in Tariffs (FFT) | Long range contracts with fixed and favorable tariffs | Easy for investors to obtain low cost loans | Dominating scheme in Europe. |
| Trading of Green Certificates (TGC) or Renewable Portofolio Standards (RPS) | Driven by official quota of green electricity | Uncertain price of green certificates reduces incitement for investors | The scheme is used by a few countries in Europe and in several states in US |
| Tender Schemes (TS) | Contracts are given to offers with lowest costs | Often used by governments in early phases of development | Used by many countries in Europe and states in the US |

# 8. ENVIRONMENTAL ISSUES

The main environmental advantages and impacts are listed in Table 8.1. During their operational life wind turbines produce between 40 and 80 times the amount of energy used for their construction.

Table 8.2 shows the comparative noise levels from different sources.

*Table 8.1: Environmental advantages and impacts of wind turbines.*

| Type | Characteristics | Remarks | Conclusions |
|---|---|---|---|
| Global warming | Wind turbines cause reduced emissions of $CO_2$, $NO_x$, ozone and fine particles | Positive Life Cycle Analysis for wind turbines | Very efficient $CO_2$ reducing technology also compared to other Renewable energy technologies |
| Health | Wind turbines do not emit $NO_x$, ozone, particles and other substances. Uncertainty related to low frequency noise. | In contrast, air pollution from fossil fuels causes increasingly allergy and respiratory deceases | Wind turbines do not create any significant health problems. |
| Visual pollution | Large turbines are visible over long distances | Careful choice of sites and good design features reduce the problem | Regulation is needed in order to reduce public opposition |
| Noise | Some noise from gearbox, blades, tower etc. Questions about low-frequency noise | Noise from modern turbines is much reduced compared to earlier models | Regulations are still needed, e.g. minimum distances from dwellings. See also Table 9.2 |
| Birds | Possible loss of habitat, breeding and feeding areas, and fatal collisions | In comparison with other causes of bird mortality the effect of wind turbines is minor | Wind farms should not be sited in bird migration routes, ref. 9.5. |
| Land use | Structure takes up less than 1% of total land needed, other activities may continue within rules and regulations | Careful planning is necessary to obtain a high energy efficiency and minimize visibility problems | With proper requirements about neighbor distance, combined use of land is possible |

*Table 8.2: Comparative noise levels from different sources (Sustainable Development Commission, 2005)*

| Source/activity | Indicative noise level dB(A) |
|---|---|
| Threshold of pain | 140 |
| Pneumatic drill at 7 m | 95 |
| Truck at 100 m | 65 |
| Busy office | 60 |
| Car at 100 m | 55 |
| Wind farm at 350 m | 35 – 45 |
| Quiet bedroom | 35 |
| Rural night-time background | 20 – 40 |

## 9. ORGANIZATIONAL SCHEMES

Organizational schemes for promotion of wind power have varied among countries. Some of the common schemes are listed in Table 9.1.

Table 9.1: Organizational schemes for promotion of wind power

| Type | Characteristics | Remarks | Conclusions |
|------|-----------------|---------|-------------|
| Individually owned turbines | Modern phase of wind power started in 1970s, especially in Denmark | The pioneering wind development in Denmark was supported by small and low cost turbines owned by individuals | Individual ownership of local wind turbines reduces opposition to wind power |
| Co-operative ownership | A joint investment from local households and farms in one or more turbines | Co-operative ownership has permitted local households to join investments in larger and more costly turbines | Co-operative ownership of local wind turbines reduces opposition to wind power. |
| Wind power developers | Electric utilities and private companies have acted as developers in relation to wind farms | With the introduction of large wind farms the need for developers with a strong capital basis is increasing | In order to avoid local opposition, wind power developers should co-operate with local interests and secure local economic advantages |

## 10. REFERENCES

BTM Consult ApS (March 2007). Report "International Wind Energy Development: World Market Update 2006". Denmark. ISBN 978-87-991869-0-7.

Cavallo, A.J., Hock, S.M. and Smith, D.R. (1993). "Wind Energy: Technology and Economics," Chapter 3 in

"Renewable Energy," Johansson, T.B. Kelly, H., Reddy, A.K.N. and Williams, R.H. (Eds.), Island Press, Washington, D.C.

Centre for Renewable Energy Technologies/ETSU (October 2000). Report "Electricity from Offshore Wind." Danish Energy Agency/IEA CADDET, Harwell, UK.

Dubois, M.R., Polinder, H. and Fereira, J.A. (2000). "Comparison of Generator Topologies for Direct-Drive Wind Turbines." In Proceedings of the IEEE Nordic Workshop on Power and Industrial Electronics, AAlborg, Denmark, pp. 22-26.

EEA Analysis of Ozone and Fine Particle Mortality (2005). Report "The European Environment, State and Outlook 2005", European Environment Agency (EEA).

Erickson, W.P., Strickland, M.D., Johnson, G.D. and Kern, J.W. (2000) Examples to Statistical Methods to Assess Risks of Impacts to Birds from Wind Plants. In Proceedings of National Avian-Wind Power Planning Meeting III, San Diego, California, pp. 172-182.

Erickson, W.P., G. D. Johnson, D. P. Young, Jr., M. D. Strickland, R.E. Good, M. Bourassa, K. Bay (2002). "Synthesis and Comparison of Baseline Avian and Bat Use, Raptor Nesting and Mortality Information from Proposed and Existing Wind Developments". Publication prepared for Bonneville Power Administration, Portland, Oregon, USA. and

Erickson, W.P. (June 2002). "Summary of Anthropogenic Causes of Bird Mortality", 2002 Wind Power Conference, Portland, OR, USA.

European Commission (2006). "Technology Focus, FP6 Projects," *Renewable Energy Newsletter*, Issue 5, Brussels, Belgium. May.

European Wind Energy Association (EWEA) and Greenpeace (2004). <u>Wind Force 12: A Blueprint to Achieve 12% of the World's Electricity from Wind Power by 2020</u>. May.

Frandsen, S., Barthelmie, R., Rathmann, O., Jørgensen, H.E., Badger, J., Hansen, K., Ott, S., Rethore, P.-E., Larsen, S.E. and Jensen, L.E. (2007). "The Shadow Effect of Large Wind Farms: Measurements, Data Analysis and Modeling," *Summary Report Risø DTU Risø-R-1615(EN),* Roskilde, Denmark. October.

Frandsen, S., Morthorst, P.E., Bonefeld, J. and Noppenau, H. (2004). <u>Offshore Wind Power: Easing a Renewable Technology out of Adolescence</u>, Summary Report, Roskilde, Denmark.

Grubb, M.J. and Meyer, N.I. (1993). "Wind Energy: Resources, Systems and Regional Strategies," Chapter 4 in <u>Renewable Energy</u>, Johansson, T.B., Kelly, H., Reddy, A.K.N. and Williams, R.H. (Eds.), Island Press, Washington, D.C.

Gryning, S.-E., Batchvarova, E., Brümmer, B. Jørgensen, H.E. and Larsen, S.E. (2007). "On the Extension of the Wind Profile over Homogeneous Terrain beyond the Surface Boundary Layer." *Boundary-Layer Meteorology*, 124, 251-268.

Hansen, A.D., and Hansen, L.H., (2007). "Wind Turbine Concepts Market Penetration over Ten Years (1995 to 2004)," *Wind Energy*, 10, pp 81-97.

Hansen, A.D. (2004). "Generators and Power Electronics for Wind Turbines." Chapter 4 in Wind Power in Power Systems, John Wiley & Sons, Ltd., 24 p.

Hansen, J.C., Mortensen, N.G., Badger, J., Clausen, N.-E. and Hummelshøj, P. (2007). "Opportunities for Wind Resource Assessment using both Numerical and Observational Wind Atlases: Modeling, Verification and Application." In Proceedings of Wind Power Shanghai 2007, Shanghai (CN), 1-3 November, 320-330.

Hansen, L.H., Helle, L., Blaabjerg, F. Ritchie, E. Munk-Nielsen, S. Bindner, H., Sørensen, P. and Bak-Jensen, B. (2001). "Conceptual Survey of Generators and Power Electronics for Wind Turbines," Summary Report Risø-R-1205 (EN), Roskilde, Denmark.

International Energy Agency (June 2006). "IEA Wind Energy Annual Report 2005". Executive Committee for the Implementing Agreement for Co-operation in the Research, Development, and Deployment of Wind Energy Systems of the International Energy Agency. ISBN 0-9786383-0-1.

Landberg, L., Mortensen, N.G., Dellwik, E., Badger, J., Corbett, J.-F., Rathmann, O. and Myllerup, L., (2006). "Long-Term (1-20 years) Prediction of Wind Resources (WASP)." In Stathopoulos, T., van Beeck, J.P.A.J. and Buchlin, J.-M. (Eds.), Introduction to Wind Technology, Rhode-Saint-Genèse (BE), 20-24 Feb 2006. von Karman Institute for Fluid Dynamics Lecture Series 2006-02) 1-21.

Larsson, Å. (2000). The Power Quality of Wind Turbines, Ph.D. Dissertation, Chalmers University of Technology, Göteborg, Sweden.

LCA Emission Analysis (1997). Report Danish Wind Energy Industry Association's "Wind Power Note" no. 16, 1997.

Meyer, N.I. (2007). "Learning from Wind Energy Policy in the EU: Lessons from Denmark, Sweden and Spain," *European Environment,* 17, pp. 347-362.

Meyer, N.I. (2004). "Development of Danish Wind Power Market," *Energy and Environment"* 15, No. 4, 657-673.

Meyer, N.I. (2003). "European Schemes for Promoting Renewables in Liberalized Markets," *Energy Policy* 31, 665-676.

Mortensen, N.G. and Nielsen, P. (1999). "New Map of Wind Resources in Denmark," *Energinyt* (in Danish) 1, 14-15.

Mortensen, N.G., Hansen, J.C., Badger, J., Jørgensen, B.H., Hasager, C.B., Youssef, L.G. Said, U.S., Abd El-Salam Moussa, A., Mahmoud, M.A., El Sayed Yousef, A., Awad, A. M., Abd-El Raheem Ahmed, M., Sayed, M.A.M., Hussein Korany, M. and Abd-El Baky Tarad, M. (2005). <u>Wind Atlas for Egypt, Measurements and Modelling 1991-2005.</u> New and Renewable Energy Authority, Egyptian Meteorological Authority and Risø National Laboratory. ISBN 87-550-3493-4. 258 pp.

Nielsen, P. (2002). "Case Studies Calculating Wind Farm Production." Report from *Energi-og Miljødata,* Aalborg, Denmark. Available in PDF format downloadable at http://www.emd.dk, 346 pgs.

Norwegian Veritas and Risø National Laboratory (2002). Guidelines for Design of Wind Turbines. (Second Edition). Copenhagen, Denmark

Petersen, E.L., Mortensen, N.G., Landberg, L. Højstrup, J. and Frank, H.P. (1998). "Wind Power Meteorology. Part I: Climate and Turbulence." *Wind Energy,* 1, 2-22.

Petersen, E.L., Mortensen, N.G., Landberg, L. Højstrup, J. and Frank, H.P. (1998a). "Wind Power Meteorology. Part II: Siting and Models." *Wind Energy,* 1, 55-72.

Rasmussen, B. and Øster, F. (1990). "Power Production from the Wind", Wind Energy Research and Technological Development in Denmark, Øster, F. and Andersen, H.M. (eds.). Ministry of Energy, Danish Energy Agency, Copenhagen, Denmark, ISBN 87-503-8305-1, pp. 7-11.

SEPA (2003). Noise annoyance from wind turbines: A review. Swedish Environmental Proection Agency, Repot 5308.

Sørensen P., Cutululis N.C., Lund T., Hansen A.D., Sørensen, T., Hjerrild J., Donovan M.H., Christensen L. and Nielsen, H.K. (2007). "Power Quality Issues on Wind Power Installations in Denmark." In Proceedings of the IEEE-PES Conference. Tampa, Florida, USA.

Sørensen, P., Meibom, P., Gehrke, O. and Østergaard J. (2006). "Technical Challenges to Energy Systems Operation and Markets," *Summary Report Risø Energy Report 5*, Risø National Laboratory and Danish Technical University, Roskilde, Denmark. October.

Sørensen, P., Hansen, A.D. and Rosas. P.A.C. (2002). "Wind Models for Simulation of Power Fluctuations from Wind Farms," *J. Wind Eng. Ind. Aerodyn,* 90, 1381-1402.

Sustainable Development Commission (2005). Report "Wind Power in the UK," UK.

Thøgersen, P. and Blaabjerg, F. (2000). "Adjustable Speed Drives in the Next Decade. The Next Steps in Industry and Academia," In Proceedings of Power Conversion Intelligent Motion (PCIM) Conference, Boston, MA, USA, pp. 95-104.

Troen, I. and Petersen, E.L. (1989). European Wind Atlas. Risø National Laboratory, Roskilde, Denmark ISBN 87-550-1482-8. 656 pages.

*Wind Directions* (2007), January/February. European Wind Energy Association (EWEA) Bimonthly Magazine January/February 2007, Volume 26, no. 2.

World Meteorological Organization (WMO) (1981). "Meteorological Aspects of the Utilization of Wind as an Energy Source," *Technical Note 175*, Geneva, Switzerland.

**ISES**

International
Solar Energy
Society

Villa Tannheim
Wiesentalstr. 50
79115 Freiburg
Germany
www.ises.org

ISBN 978-1-84407-539-3

9 781844 075393

http://www.earthscan.co.uk
ISBN: 9781844075393

**ISES**

International
Solar Energy
Society

ttp://join.ises.org

ISES Vision:
Rapid Transition
to a
Renewable Energy
World

ISES is a global alliance with a vision: Rapid Transition to a Renewable Energy World

Since 1954 ISES has been serving the needs of the renewable energy community. The goals of ISES include:

## ISES

## International Solar Energy Society

• Encourage the use of renewable energy globally through appropriate technology, scientific excellence, social responsibility and global communication;

• Realise a global community of industry, individuals and institutions in support of renewable energy technologies;

• Support the development and the science of solar energy.

**Join today and be a part of helping ISES achieve this vision!**
**http://join.ises.org**

International Solar Energy Society
Wiesentalstr. 50
79115 Freiburg
Germany

Tel: +49 761 45906 0
Fax: +49 761 45906 99
Email: hq@ises.org
Web: www.ises.org